AF715311

# Ecohouse - A Holistic Approach to Sustainable Living

Rogerio Cietto

Published by Rogerio Cietto, 2012.

# Ecohouse – A Holistic Approach to Sustainable Living

Published by Rogerio Paiva Cietto at Smashwords

Copyright 2016 Rogerio Paiva Cietto

Thank you for downloading this book. You are welcome to share it with your friends. This book may be reproduced, copied and distributed for non-commercial purposes, provided the book remains in its complete original form. If you enjoyed this book, please return to Smashwords.com to discover other works by this author. Thank you for your support.

# Index


**Introduction**

- **Shelter**
- **Water**
- **Energy**
- **Food**

# Introduction

*God blessed them and said "Be fruitful and increase in number; fill the earth and subdue it. Rule over the fish of the sea and the birds of the air and every creature that crawls upon the earth" Genesis 1, 28.* (so every man is called to rule over the sea, the air and the land, and that's why we all have some SEAL's and similar special forces' spirit on us).

The word eco means nature (in Greek *oikos* means house), so an Ecohouse would be a place that respects the environment (literally, a house inside a house). It would be easy for all of us to be ecologically friendly, in complete symbiosis with nature: we just need to live the way we did 10.000 years ago, hunting, planting, and living in shelters with no water or energy...

It would not be wise, since many of us would die from starvation, diseases or disasters. But the way our ancestors lived may help us understand why we need to rethink our relation with nature. Our ancestors fought for their individual survival on a daily basis, and those with technology and resources (virtually unlimited at that time) had a step ahead.

Today we fight for the same survival, but based on the benefits that we have today, and some of us are way ahead because of access to technology and resources (clearly limited when faced to the world population today).

Before explaining the number of ideas gathered in this work, it's necessary to understand Sustainability and the concepts and principles that inspired this work.

Sustainability is the ability to live respecting three factors:

- environmental (good for the planet and its resources);
- social (good for other people, be they near or far);
- economical (accessible to everyone to purchase and maintain).

Pope Francis clearly exposes this interrelation in the encyclical letter *Laudato Sì*: "139. When we speak of the '**environment**', what we really mean is a **relationship existing between nature and the society which lives in it**. Nature cannot be regarded as something separate from ourselves or as a mere setting in which we live. We are part of nature, included in it and thus in constant interaction with it. Recognizing the reasons why a given area is polluted requires a study of the workings of society, its economy, its behavior patterns, and the ways it grasps reality. Given the scale of change, it is no longer possible to find a specific, discrete answer for each part of the problem. It is essential to seek comprehensive solutions which consider the interactions within natural systems themselves and with social systems. **We are faced not with two separate crises, one environmental and the other social, but rather with one complex crisis which is both social and environmental**. Strategies for a solution demand an integrated approach to combating poverty, restoring dignity to the excluded, and at the same time protecting nature."

Therefore, the ideas described are aimed to those with few natural resources, social problems and financial restrictions. If you are not in one of these categories, you can use them easily, or even improve them, according to your needs. Anything you save from nature will be good to others.

Due to severe environmental conditions some people live today, like droughts, exposure to biohazards or starvation, building an ecohouse improves significantly their chance of survival and, more important, can restore their dignity.

What does an ecohouse need to have? At least it has to allow the possibility of achieving the main necessities of any human being, which are:

- protective shelter;
- drinkable water;
- energy resources;
- nutritious food;
- breathable air;

- being part of a family and a community;
- believing in transcendence and salvation.

The main needs of the human being will never change, but the methods to achieve them varies. That's why some of the ideas will work very well in some places, and not so well in others. Feel free to adapt our suggestions to your own specific needs. Adapting is essential for the survival of any species.

The five first needs are physical, and the last two are spiritual. Our work will show how to have an environmentally friendly shelter with energy, water and food. Clean air is found only far from cities, where there are forests. The spiritual needs can be found only inside a family and inside yourself (more information in the book *Man's search for Meaning,* from Viktor Frankl). We advise you to seek which doctrine is true salvation, and avoid false prophets and promises.

Although a family and a community are not made of solid materials, their links (that is, father-mother-children, which will never change) are needed for any person who seeks peace, moral values, spiritual fulfillment and happiness. Also, each of its residents must be protected from hopelessness when things don't happen according to the plan. In addition, every human being needs to believe in something beyond him or herself, and everything that happens on their lives are temporary, but salvation is forever.

Certainly, a well-structured family will benefit from a well-structured house, where parents will be able to live in peace and raise their children. In addition, children will learn easily all concepts related to natural sciences, because they will experience it every day.

Finally, we found out that when implemented many excellent ideas just don't work. Ex: rainwater harness in the Brazilian northeast, or shelters in sub-Saharan Africa pose unique challenges. We will work in a worst-case scenario, so we:

- Don't rely on any government support. This way your sustainability won't depend on politicians, lobbyists, government funds or the like.

Nevertheless, cooperative work is always important, and working in groups is part of our social life;

- Believe that living is not just surviving, but living with dignity and security. So, having some room space and solid walls is not superfluous. Ask yourself: "Would you live in a house with thin walls and room for two only?";

- Believe that simple ideas and solutions cost less. A solar panel is not accessible to everybody in the world, and even if they were, who would pay for it? It's good to respect the environment, but how much would you pay for it? It's not reasonable to pay 50% more for an organic apple.

A very simple concept of sustainability is taking from nature the same that you give back. Taking is easy, and inevitable (food, water, energy, natural resources) though if you take less, you would need to give less to be sustainable.

Ways not to take much from nature:

- don't waste food, even if it's not proper for consumption (don't eat it, though);
- recycle and reuse water, especially when there's not much rain;
- produce the least possible amount of carbon dioxide, methane and other greenhouse gases;
- use renewable energy (air, water, solar, etc.);
- don't waste materials used for construction.

Ways to give back what nature provided you:

- plant trees easily adaptable to the climate, so as to produce oxygen and absorb carbon dioxide;
- use as many reclaimed objects (plastic, metal, etc.) as possible;
- try as much as possible to help with nature's cycles (water cycle is part of the project, and you can find easily that food cycle can be helped too).

The picture below is a simple way of showing how we may keep our relationship with the environment (you'll see that the idea of this book started as a school project). As long as the flow of resources (red arrows) is going, the cycles are being respected, and the person can affirm he has a sustainable life. If the cycle is not flowing properly (evaporation, trees photosynthesis, water absorption in the soil), some parts of it are overloaded for the benefit of the person.

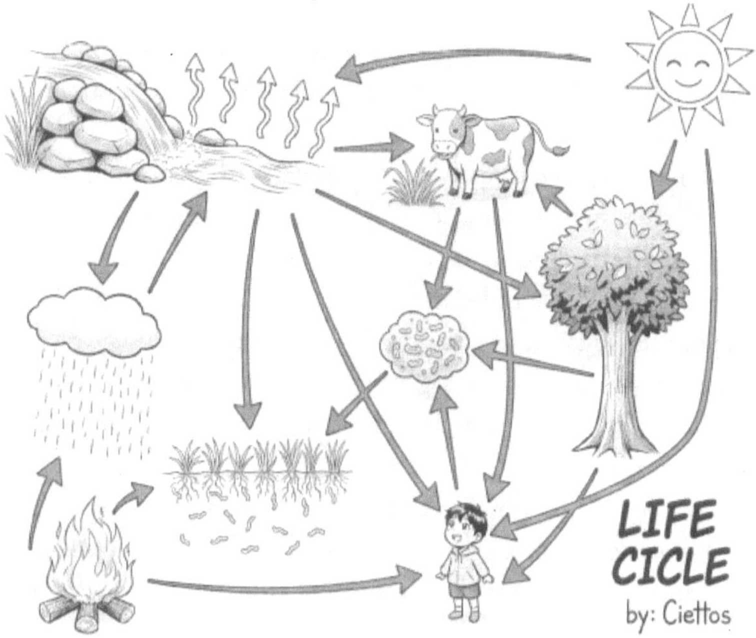


Sustainability is a process, and the best way to put the ideas into practice is gradually. So don't become radical, don't throw away your new car or gas stove, until they are worn out (then it's time to have a car working on electricity or ethanol fuel, and an electric or biogas stove).

Nevertheless, be aware that becoming sustainable takes some time, money and work. But when you achieve it, you will be free. Not totally independent, but the government won't affect your life and your family as much.. Freedom is the objective of every human being, and working and fighting for it is worth every second, cent or sweat spent.

# Shelter

In a survival situation a person may try to find water or food first. Nonetheless, in a shelter he/she will be away from many dangers, and will be able to fulfill his/her other needs easier. Lack of an adequate place to live is the most notorious problem in poor countries, and from that comes the excessive use of other resources (more energy to cool down bad insulated houses, water polluted due to all kinds of waste sent directly to water supplies, etc.). The lack of housing is also an important cause for the disruption of families.

Finding a place to build a house in an urban area may be difficult, due to the high cost of land, but it's much safer building your ecohouse in a small safe area than a larger place with risks of flood or other natural disasters. Also, people who live in the rural area may want more space for agriculture.

We designed a house based on a 175 sq m property, based on real state costs and local legislation in many cities in poor countries. A 7 x 25 m property is usually the minimum available, and it's enough for four to five people to live in good conditions (35 to 43 sq m/person).

If your place is bigger than 7 x 25 m, use the extra space to plant trees or anything edible, but be careful when choosing the plants, some may attract poisonous animals, or the roots may grow horizontally and affect the building. Also, don't plant trees that will grow more than 8 meters and shade the roof with the solar panels.

The house may eventually be used for even more people, since in poor countries it's common that fifteen or more people live under the same roof. It's not the ideal situation, but the worst-case scenario we want to deal with.

Certainly, this space is not available anymore in New York City, Sao Paulo or Tokyo, where more than 500 people live in a building that uses 1,000 sq m. of land. This kind of living (2 sq m/person) requires a lot of energy and resources, and is far from sustainable. For those in this situation, using the water and energy ideas explained below is of utmost importance.

We also consider that the place has no energy, water or sewage services. If they do, it's very likely that they won't be sustainable nor reliable, so it's better if you think of living off the grid.

The house designed below is very simple, and you can adapt it to your needs. But remember that the water reuse and recycling use gravity to work. If possible, build the main axis on the east-west direction, and the lower roof facing north (South Hemisphere) or south (North Hemisphere) in order to heat the water used to produce energy.

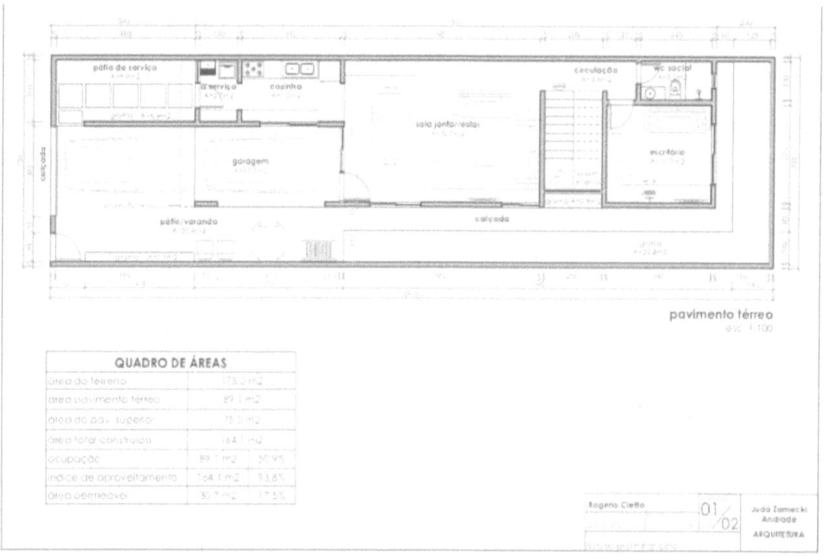

10

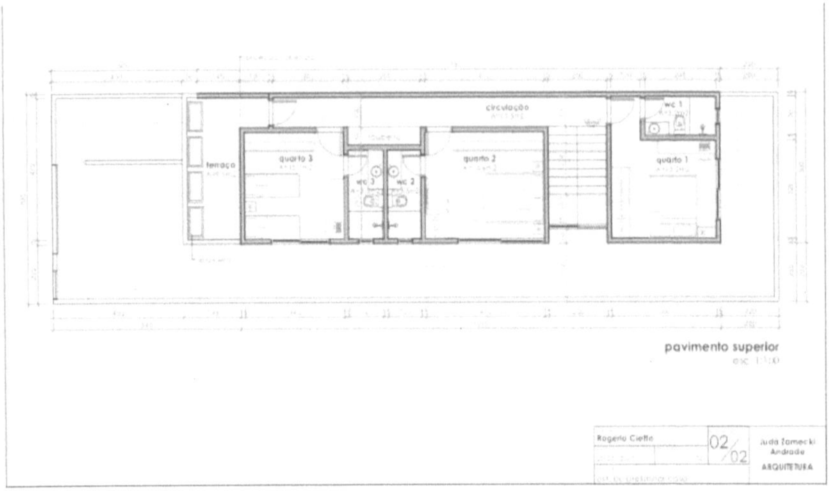

circulação
terraço
quarto 3
wc 3
wc 2
quarto 2
wc 1
quarto 1
pavimento superior
02/02
ARQUITETURA

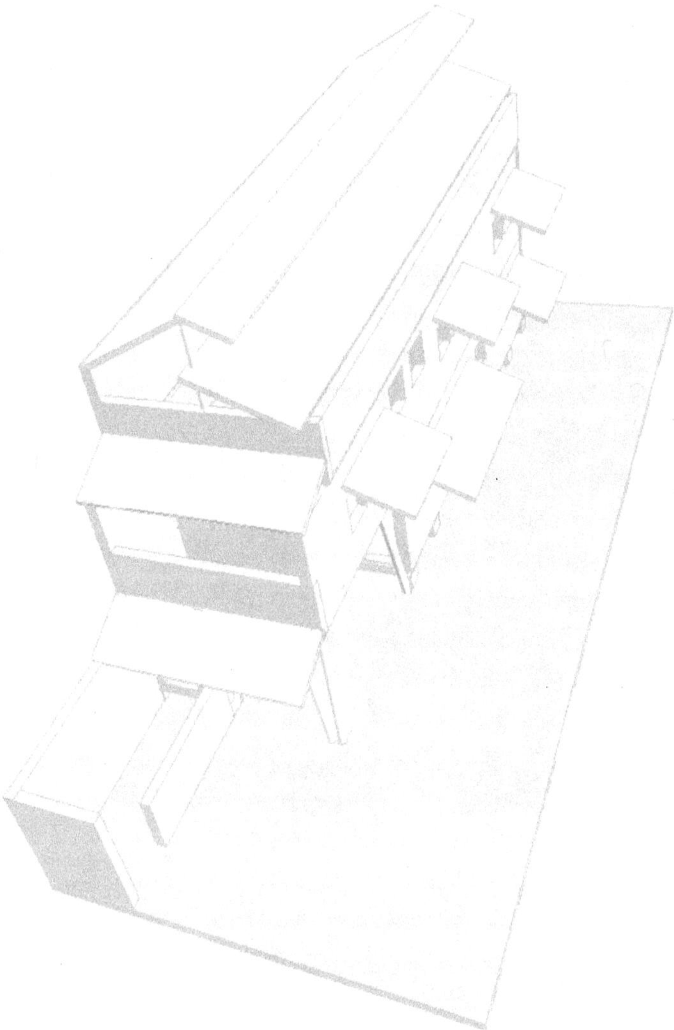

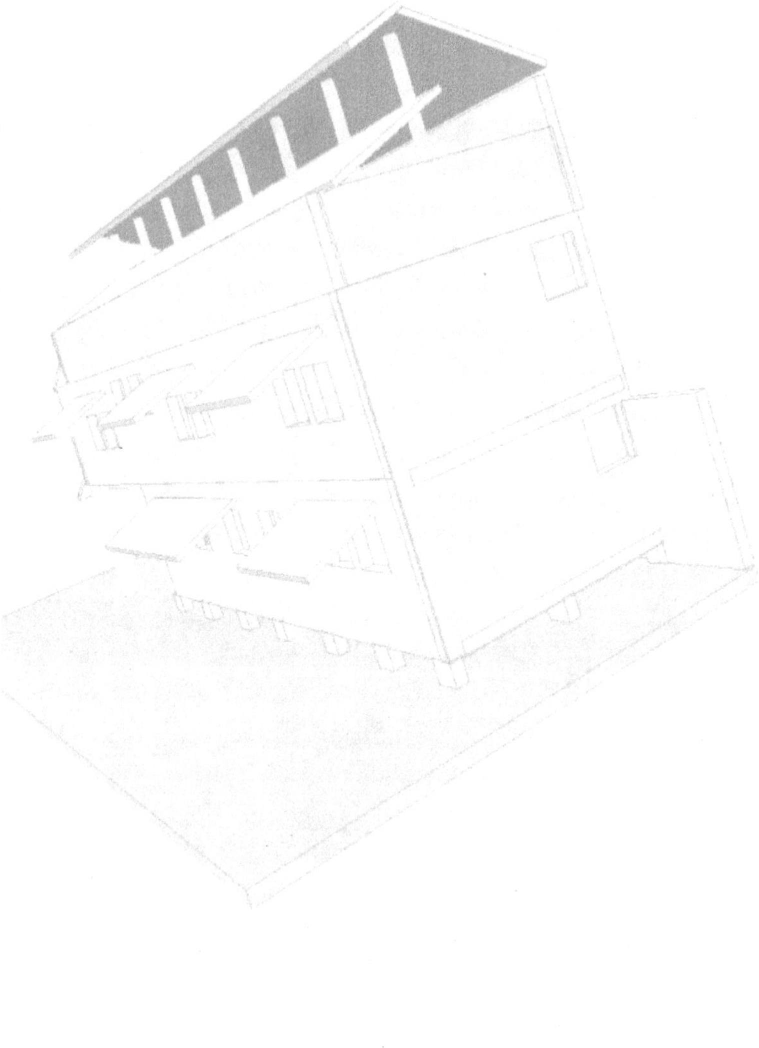

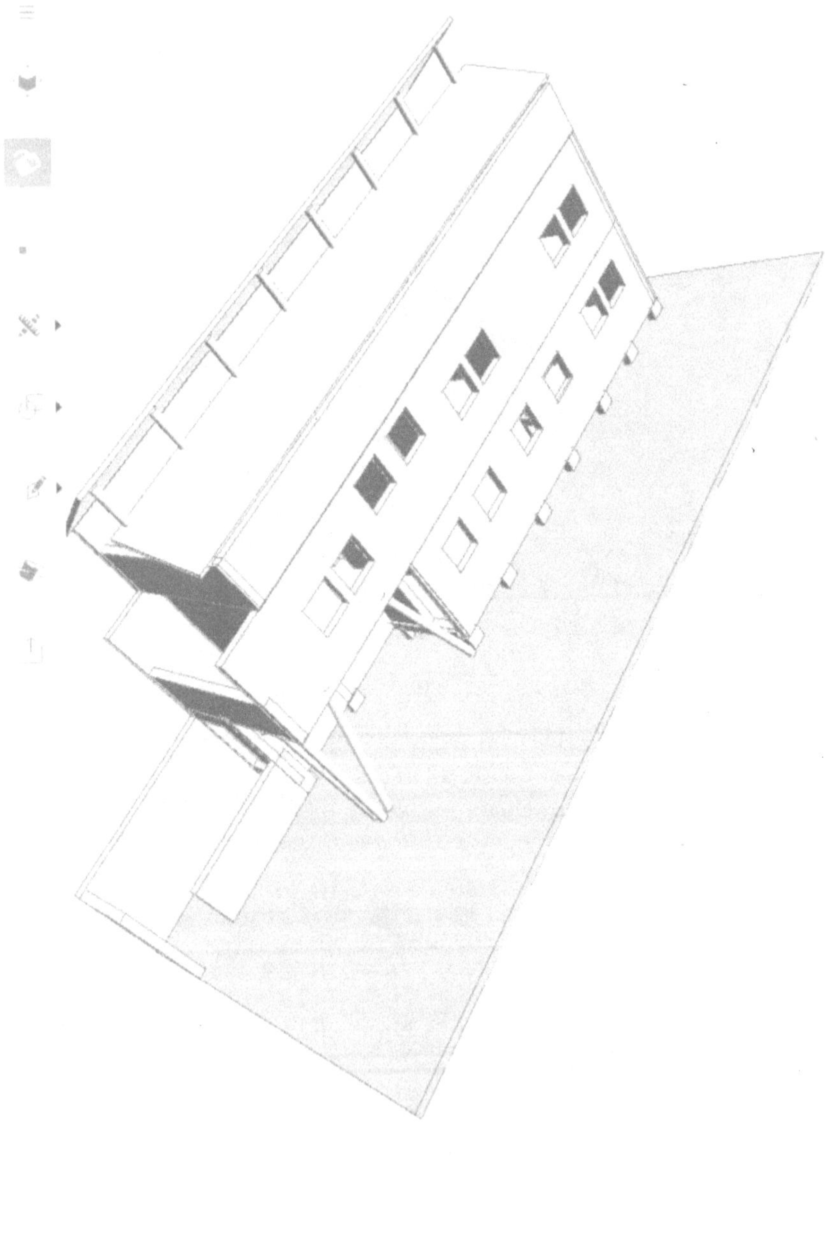

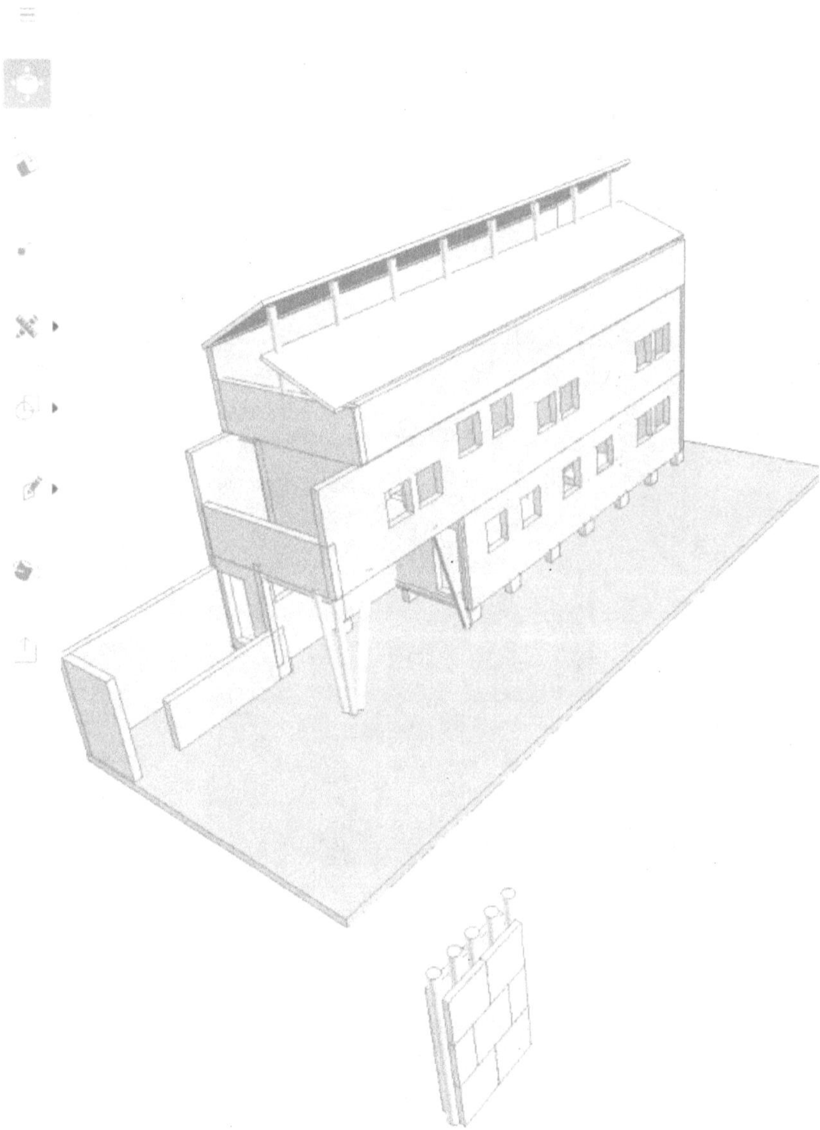

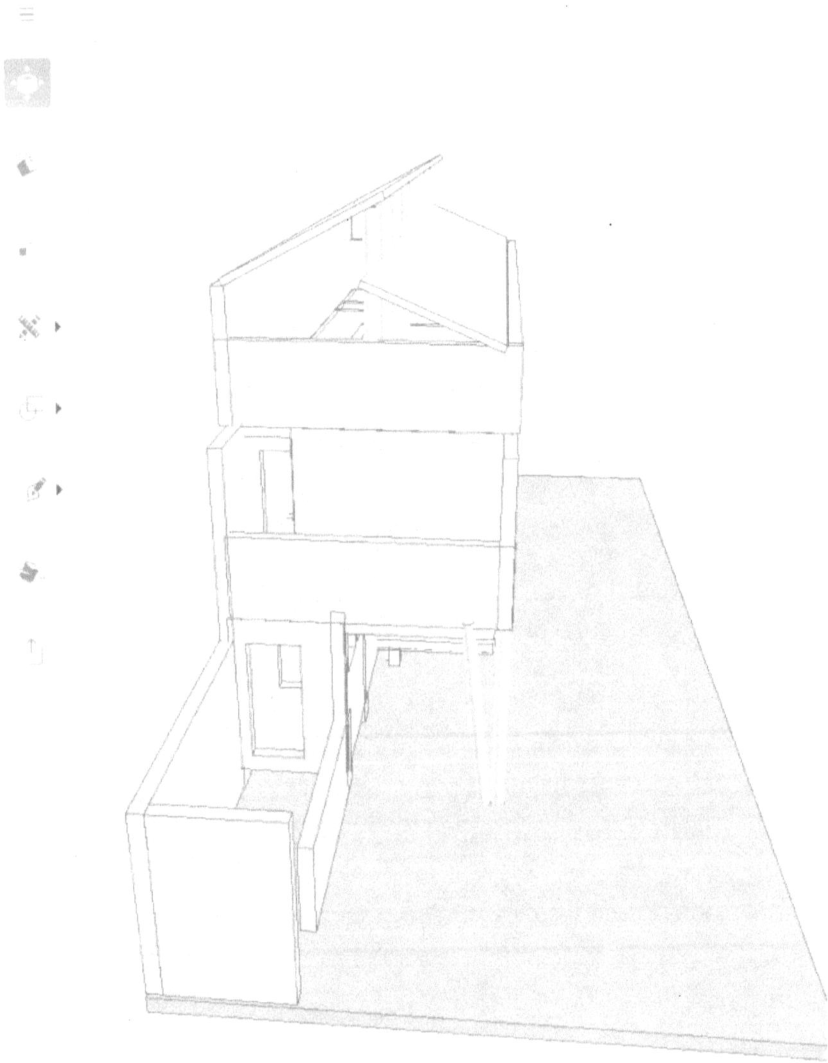

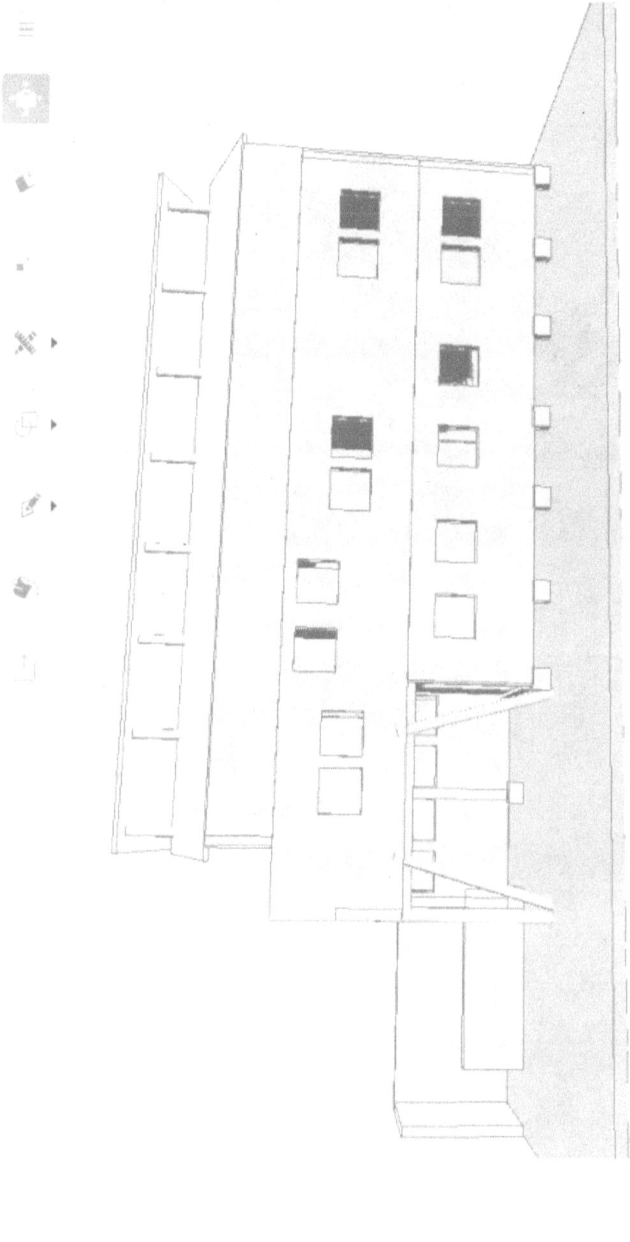

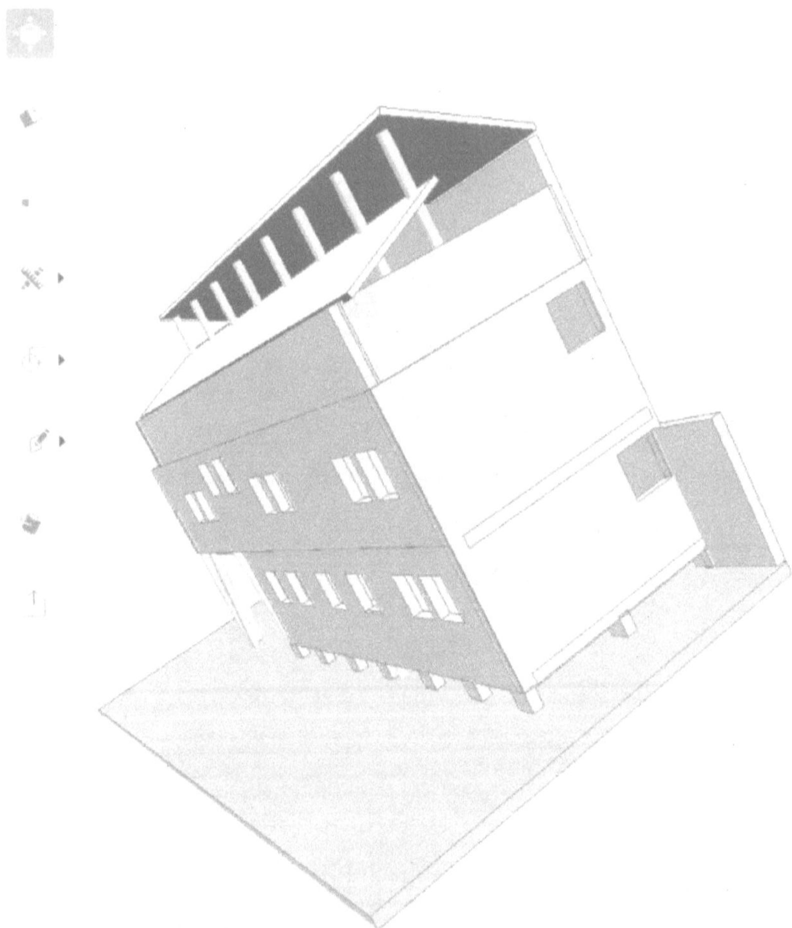

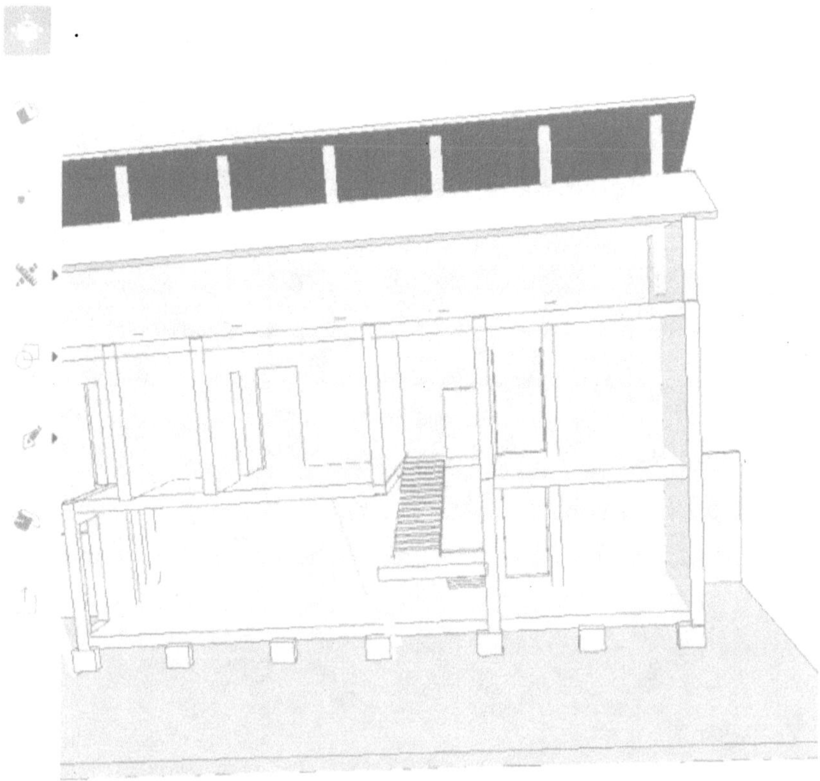

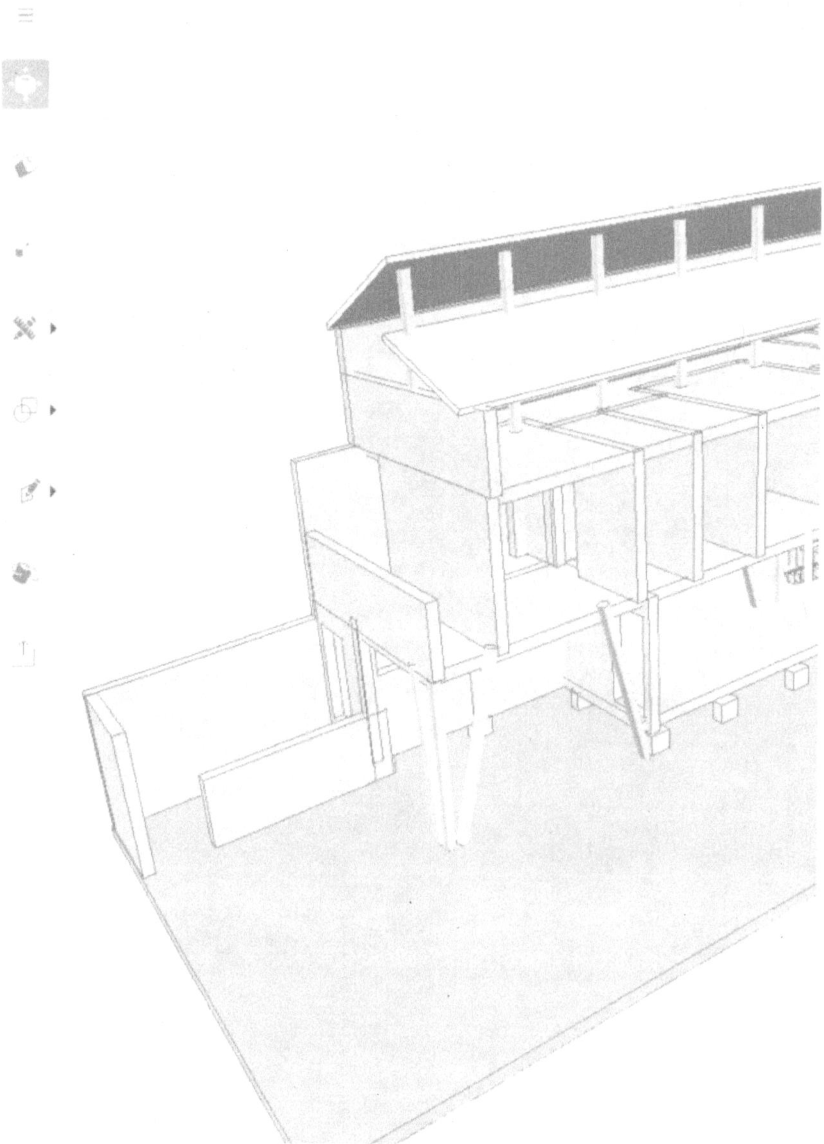

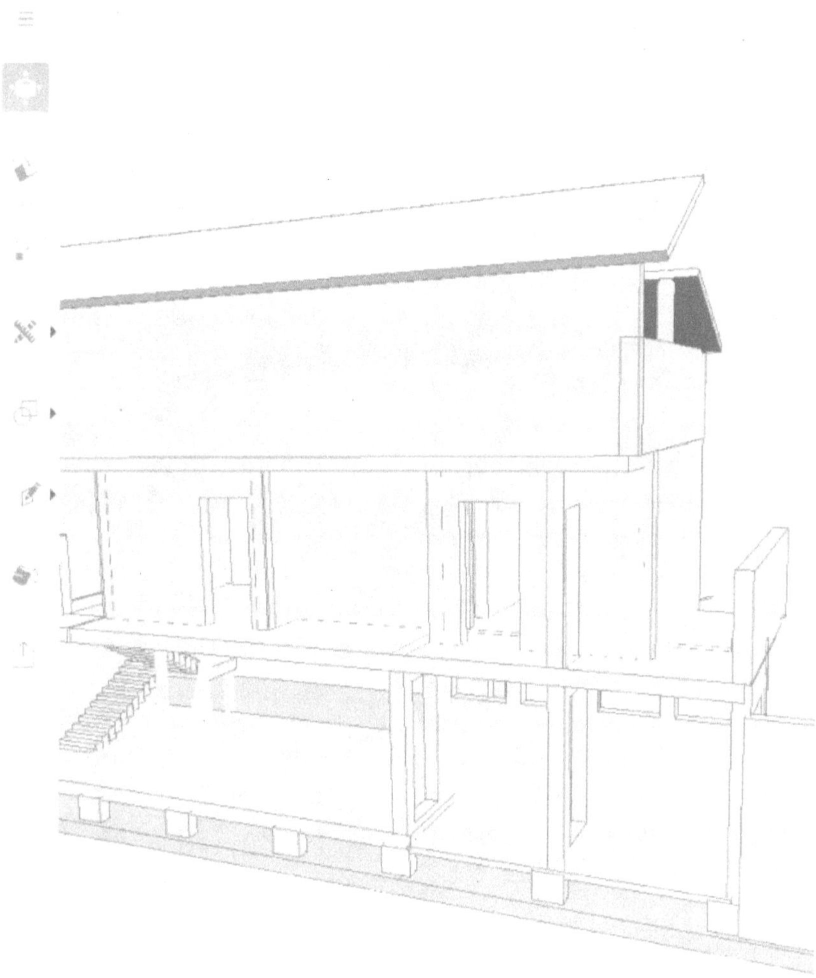

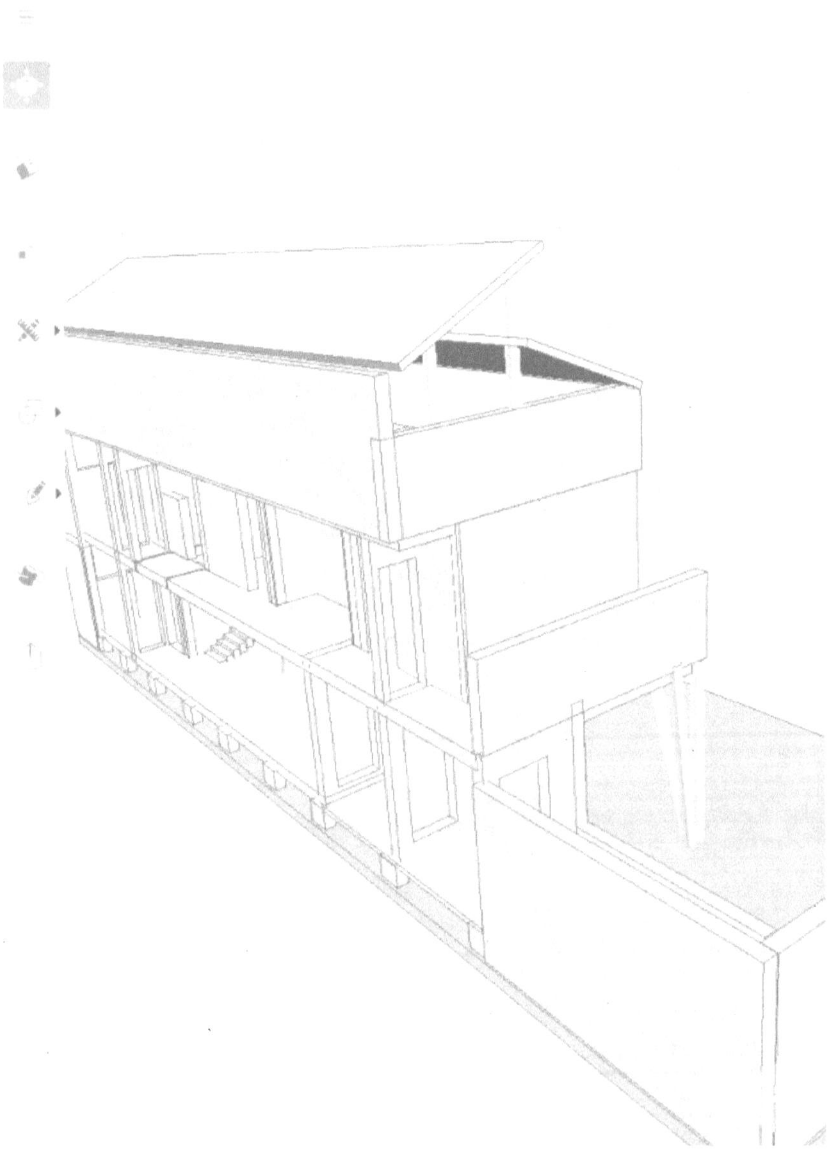

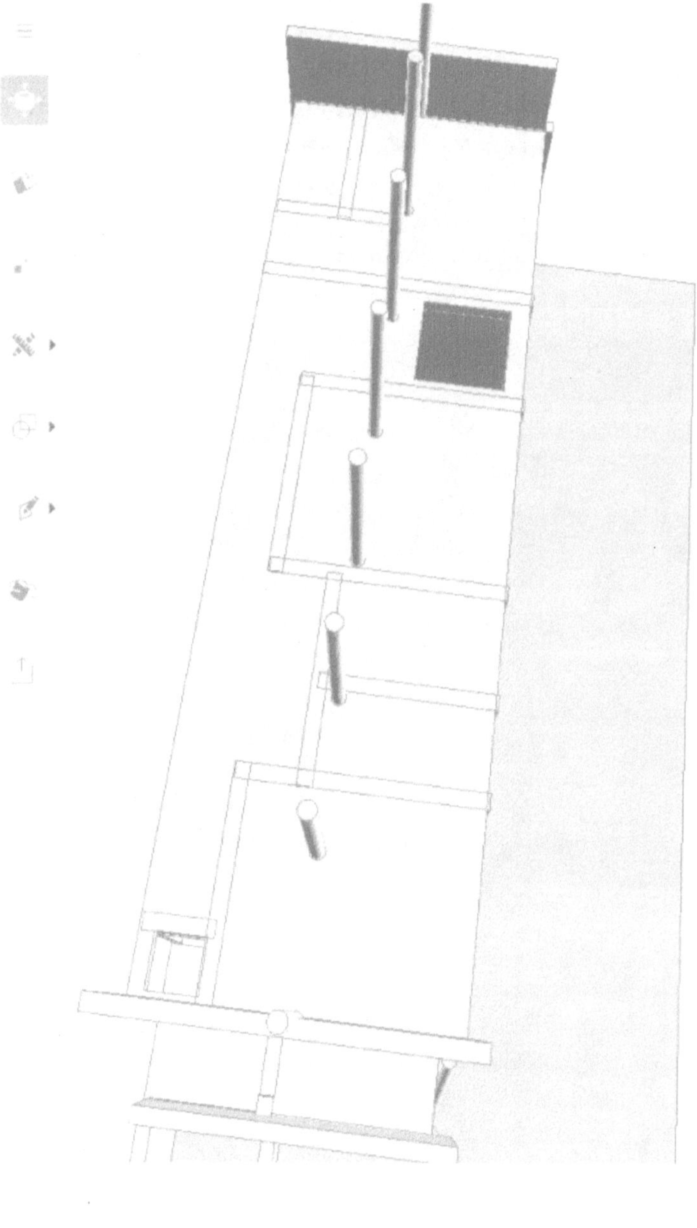

The main windows should be facing south or north depending on where you want to build, following the side of the lower roof. On the Northern

Hemisphere the windows must be facing south, and on the Southern Hemisphere the windows should be facing north. The following images shows the position of the house according to the latitude (Line of Equator in red, Tropics in green).

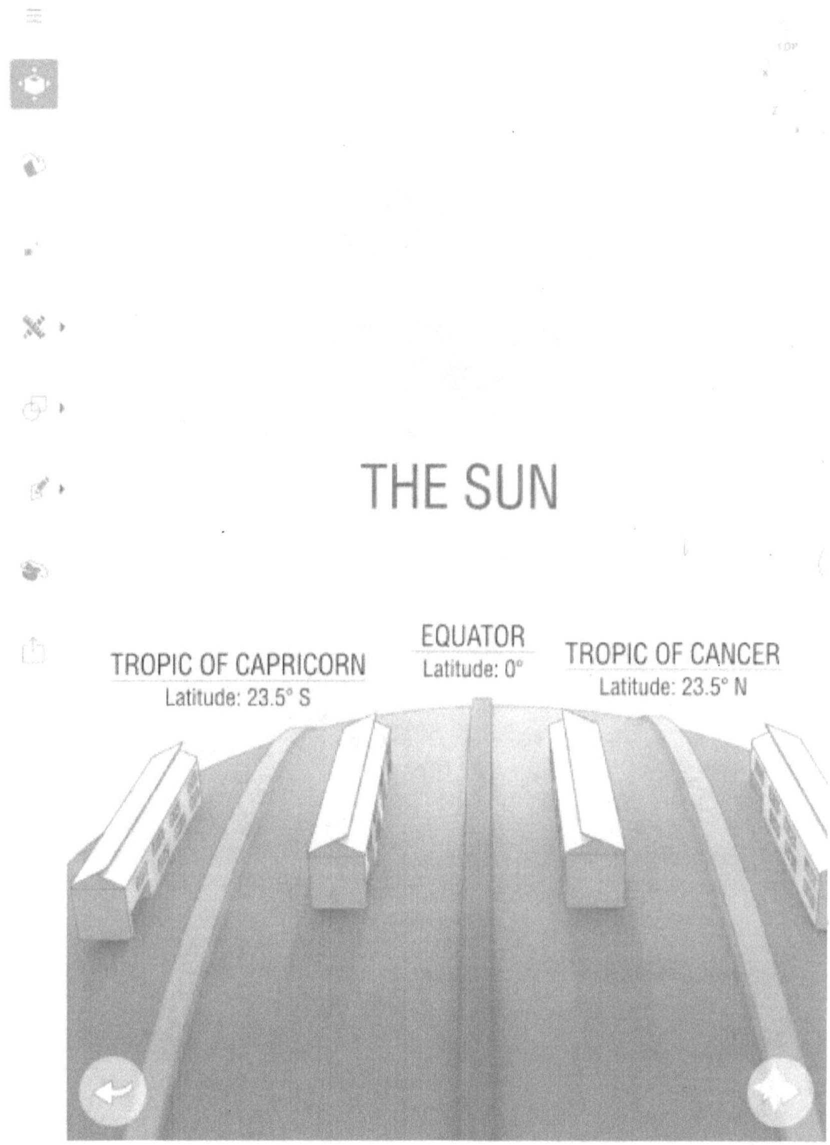
THE SUN
TROPIC OF CAPRICORN
Latitude: 23.5° S
EQUATOR
Latitude: 0°
TROPIC OF CANCER
Latitude: 23.5° N

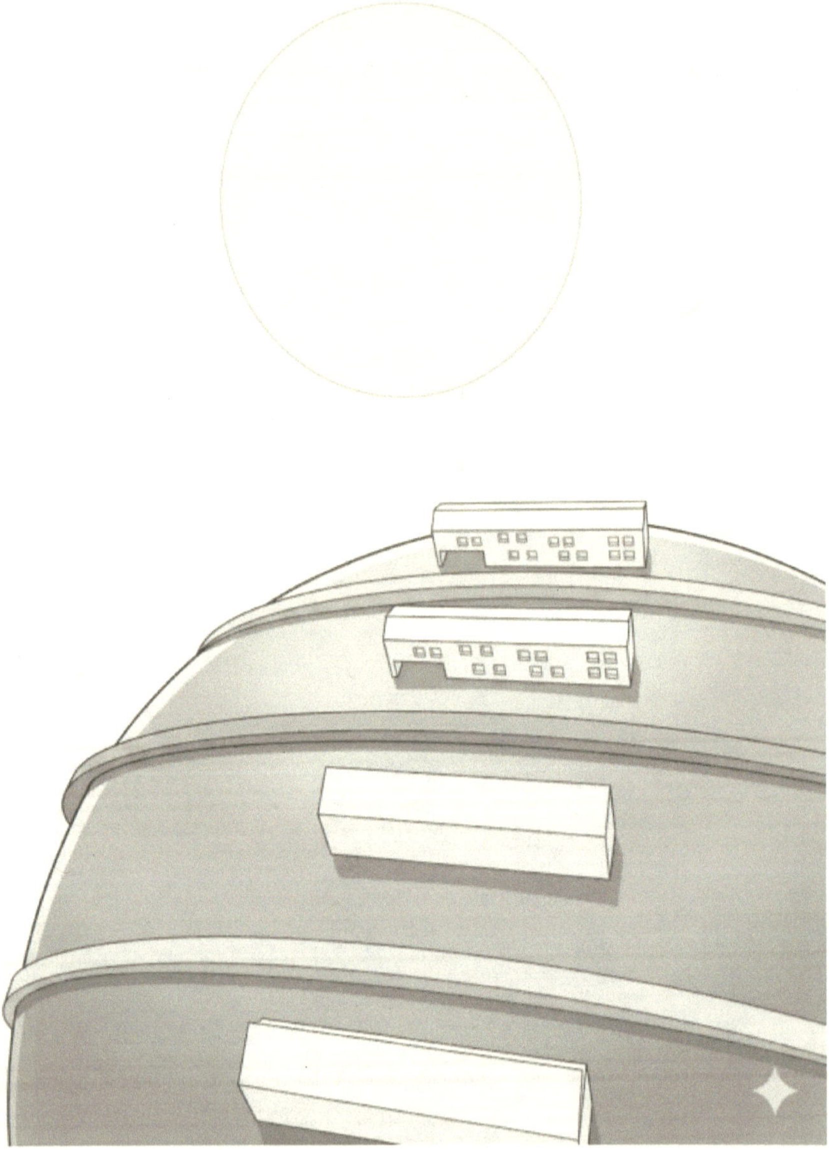

If the wall with windows is close to another wall or your neighbor, it's important to paint this wall opposite to the windows in white, in order to allow a good illumination for the house. If possible, build the wall with windows **at least six meters apart** from another wall, allowing adequate illumination and an area to plant large trees.

**First floor**: garage, living room, dining room, kitchen, laundry room, winter garden, toilet, office room, stairs for the second floor, and backyard.

Plant something aromatic at the winter garden. The winter garden and part of the stairs are open up to the roof, to provide good ventilation and some illumination.

All dry organic waste is burned in the furnace at the back of the house, creating energy to boil the water, purifying it. The vapor is used to run an electric generator on the third floor. If there's not enough burning material, use the biogas from the biodigester or an electric resistance to boil the water (you need a manual generator to run it, or use it when there's nothing else on; you can also use a bicycle frame to run the generator pedaling). When the vapor reaches the roof, it turns an electric generator, condensates and is stored to the hot water tank.

Humid organic waste (ex. leftovers) can be grounded and sent to the sink. It goes to the sewage tank in order to produce biogas and fertilize the soil. It's good to have a triturator in the kitchen sink.

Recycled materials shall be separated and reused, if possible. Bins with different colors and some discipline is all you need.

There are two tanks for the sewage: one is for the washing machine, you can use the graywater from it (actually it's blue, so you can call it blue water) to water your plants, wash your terrace, garage, etc. This served water is also a good pesticide, but it does not kill weeds.

The rest of the sewage (kitchen, toilets, bathrooms) goes to another tank. Through decantation the liquids (on top) go to another tank, situated over the furnace at the back of the house, in order to be boiled, sent up as vapor to turn an electric generator on the third floor and condensed back to be used in the hot water tank.

The remaining graywater goes to an odorless biodigester to produce biogas, which can be sent to the kitchen stove or also be sent to the furnace, to boil the served water. It's important to build the biodigester close to the kitchen (but five meters away from the stove) and the furnace at the back of the house, so that the biogas produced and the furnace are far from each other. Close to the biodigester you can use another tank to collect the bio digestate (the remaining product of bio digestion) and use it as fertilizer, or send it to the ditch.

**Second floor**: master bedroom with bathroom, stairs and two single bedrooms, each with a bathroom. The windows can have a covering to shade from sunlight during summer, and in warm climate regions the windows should be always against the sun.

**Third floor**: tanks to collect rainwater are installed on each corner, in a total of four (it's easier and cheaper to have four tanks with water ready to use than one big cistern on the ground). There are also two hot water tanks, and three energy generators. Cold water tanks can be interconnected, in order to share their capacity (capillarity). Cold water must be sent up to the hot water tank eventually, so use an electric pump with a sensor.

It's also important to respect local laws when designing your house. One common requirement is that at least 20% of the total area is not constructed (green area). The garage and backyard provide enough open areas in order to reach the 20%. The way it is shown, the project has approximately 164 sq m of constructed area (89 sq. m. of occupied land), and 30 sq m of permeable area (green area) plus a permeable terrace. If your property is bigger than 7 x 25 m, just follow the design and there will be plenty of green.

Materials used vary according to the local availability. We suppose that anywhere in the world a person can find bamboos, a very versatile and durable plant, when treated. Bamboo treatment with borax is considered today the most adequate for human use and for the environment. Another advantage of using bamboo is the growing capacity of the plant and the soil regeneration. You can also use wood, bricks or other renewable materials, but they may be more expensive.

Bamboos adequate for construction include the species *Guadua Angustifolia* and *Dendrocalamus Asper*, they are easy to plant and adapt well to tropical climates. After treatment the bamboos have to be selected according to the width of their walls. Thick walls (> 1 cm) should be used as structures, and the others divided in stripes and glued in two layers, one horizontally and one vertically, making a product called Glued Sliced Bamboo (GSB). Bamboos too thin for the GSB can be used to connect two bamboo pieces, putting it inside and fixing. Use a drilling machine and bolts in order to fix the pieces and build the structures for walls and floors. Use metal pipes or bars to connect bamboos in beams and columns.

Although the bamboo cover has a good protection against the sun and rain, it's important to apply a layer of waterproof varnish or waterproof liquid plastic cover on the buried pieces or those more exposed to the weather, so as to make all the construction last longer.

This is an example of how to assemble bamboo pieces and GSB in order to build walls and floors. All GSB are 1 m x 1 m, and bamboo pieces are 3 m x 20 cm. To install a window, you just need to unassemble (or leave open) one GLB, and reinforce the frame. For doors open two or more GSB pieces. When assembling use stronger ones for beams and columns (wall thickness 1 cm or thicker).

Advantages of this assembling: it's very easy to build, maintain and repair; installing and maintaining pipes and wires is even easier; a efficient cooling system is created, so the hot air comes up continuously in the open spaces inside the walls, leaving the inner side cool enough even in the heat of the Amazon Rainforest region. For cold climate regions you need to fill the empty spaces with clay, foam or another insulating material.

ピカ
ザワッ

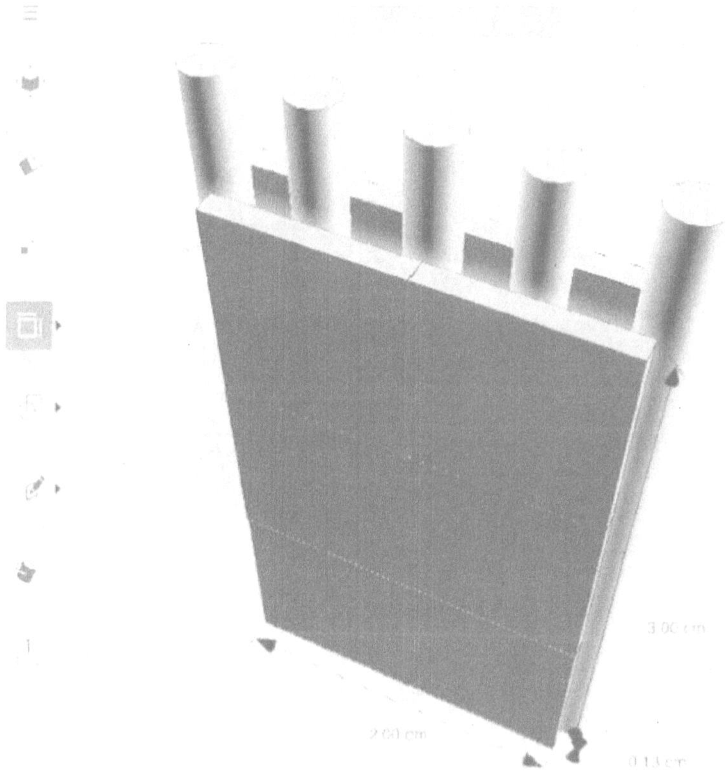
3.00 cm
2.00 cm
0.13 cm

This is... artisanal split bamboo wall...
This is... artisanal split bamboo wall...
SHIII
BAMBU-
What a unique bamboo structure!

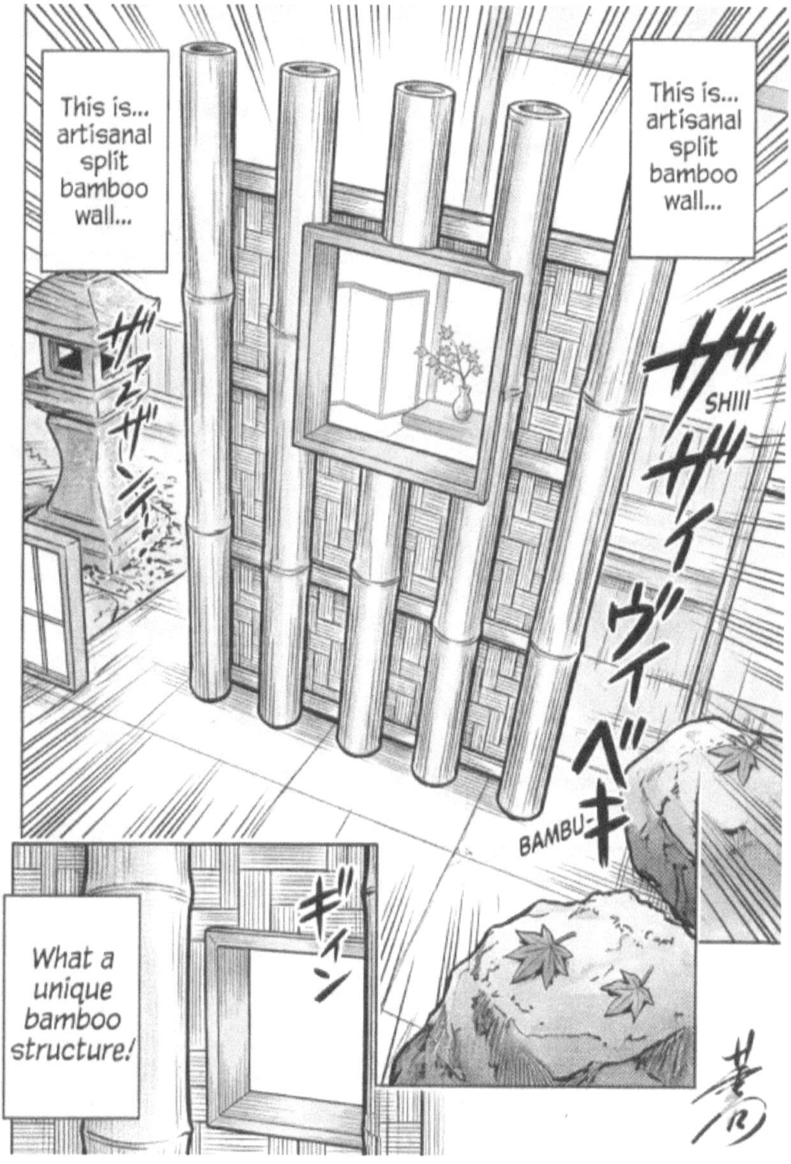
This is... artisanal split bamboo wall...
This is... artisanal split bamboo wall...
SHIII
BAMBU-
What a unique bamboo structure!

Plastics, clay or other fabrics also offer a good insulation, important to prevent heat from coming in in summer and going out in winter.

Doors and windows must also have good insulation. Don't rely on air conditioning to cool down or heat your house, just because it's cheaper to build. The cost will be much higher to keep it habitable.

# Water

Water can be found underground, but drilling the pit may be very expensive, and the water you take needs to be pumped to the top. A simpler solution is to collect rainwater, available even in dry lands like the dry savanna in the Brazilian Northeast.

Recycling your reclaimed water is another way to have this important source, and it's a must for a sustainable living in places with no sustainable sewage services available. Also, using less water will help. For example, toilets with two options: small flux and large flux.

Reclaimed water needs to be boiled or filtered before use. The more you use, the more you will need to clean. Therefore, it's wise to reuse water (from the shower to the toilet, and from the washing machine to water the plants and to clean the outside of the house).

The reclaimed water is boiled using two sources: dry organic leftovers, non-recyclable paper and biogas from the biodigester are burnt below it (it pollutes a lot less than waiting for biodegradation, and it is considered sustainable if it cannot be recycled), and also an electric heater or resistance (run by a manual energy generator, adapt it to a bicycle frame).

The energy production may be higher than consumption during the day, so this extra energy can be used to heat the served water and send the vapor up. Since the system above uses hot water to generate electricity, it's a very simple way to store energy (potential energy in the form of water pressure is stored in the hot water tanks, and become electrical energy when required).

Always use soap and detergent with biodegradable tensoactives (specially with glycerin, the bacteria at the biodigester will boost the production of biogas with it), so the remains of the water-recycling system can be used (biogas is produced in the biodigester, and sent to the stove or to boil the served water) and as a fertilizer. Organic waste must be ground and sent to the sink. Filter all water used, because chlorine will prevent the anaerobic bacteria from producing biogas.

**Never use caustic soda (sodium hydroxide) to unclog sinks and drains**, this chemical product will ruin the biodigester and require a complete cleaning of the tanks in order to generate biogas again.

If you eventually produce too much reclaimed water, you may need to send it away, so build a ditch on the backyard, put some rocks on the bottom to filter the water from the debris, and next to it plant as many water-consuming plants as possible. But use it as a last resort, only if you cannot recycle the water, or it's raining season.

In addition to using four water tanks on the third floor, or in dry places with short raining, it's important to build a cistern on the first floor, and use a water pump to send it to the water tank for use. Build it on the backyard, opened at the top, above the ditch, and use some chlore to keep it clean (like a swimming pool).

You can throw some animal or vegetal oil/fat on the sink (never mineral oil), it will be well consumed by the anaerobic bacteria, but don't send too much. You can also use it to make soap (mix it with sodium hydroxide, aka caustic soda, and a syrup with aromatic leaves or barks (bamboo, Pinetree, eucalyptus cypress, and similar are also good antibacteriological agents). Or just burn it, if you don't have time.

Laundry Powder
Softener
The water is collected ...
...and filtered
Heat
Regular
Rinse
Wash Cycle
Wash Cycle
Wash Cycle
...and ready to use for the next load!
It's a lot of work, but it's important!
Washing Bucket - Cleaned Wa

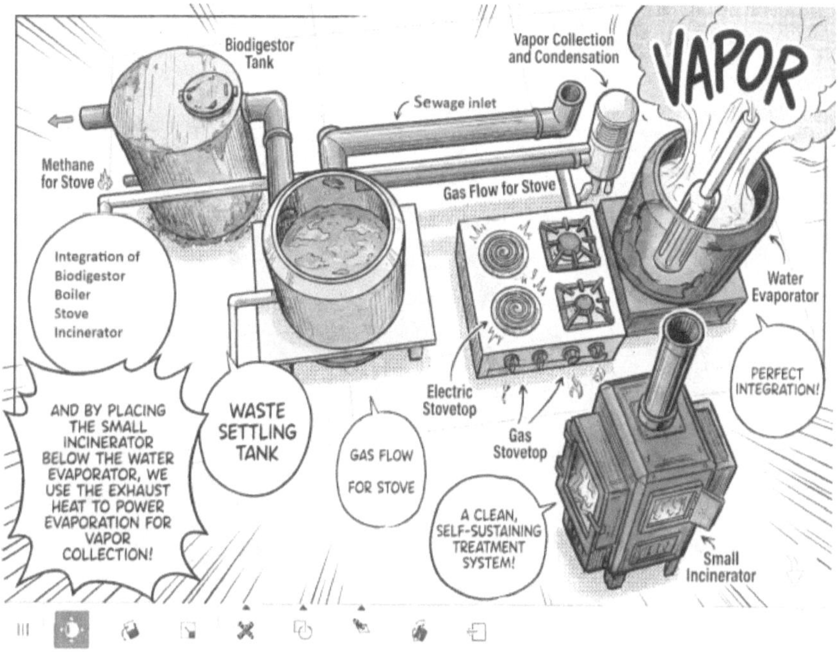
Biodigestor Tank
Sewage inlet
Vapor Collection and Condensation
VAPOR
Methane for Stove
Gas Flow for Stove
Integration of
Biodigestor
Boiler
Stove
Incinerator
Water Evaporator
PERFECT INTEGRATION!
AND BY PLACING THE SMALL INCINERATOR BELOW THE WATER EVAPORATOR, WE USE THE EXHAUST HEAT TO POWER EVAPORATION FOR VAPOR COLLECTION!
WASTE SETTLING TANK
GAS FLOW FOR STOVE
Electric Stovetop
Gas Stovetop
A CLEAN, SELF-SUSTAINING TREATMENT SYSTEM!
Small Incinerator

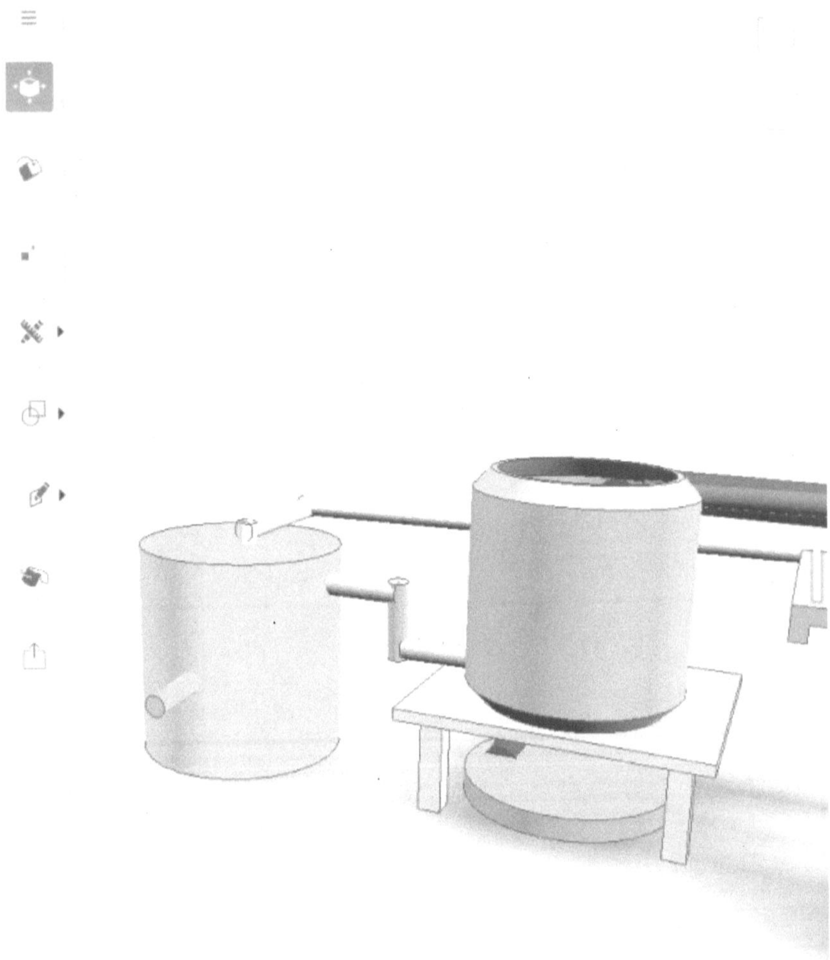

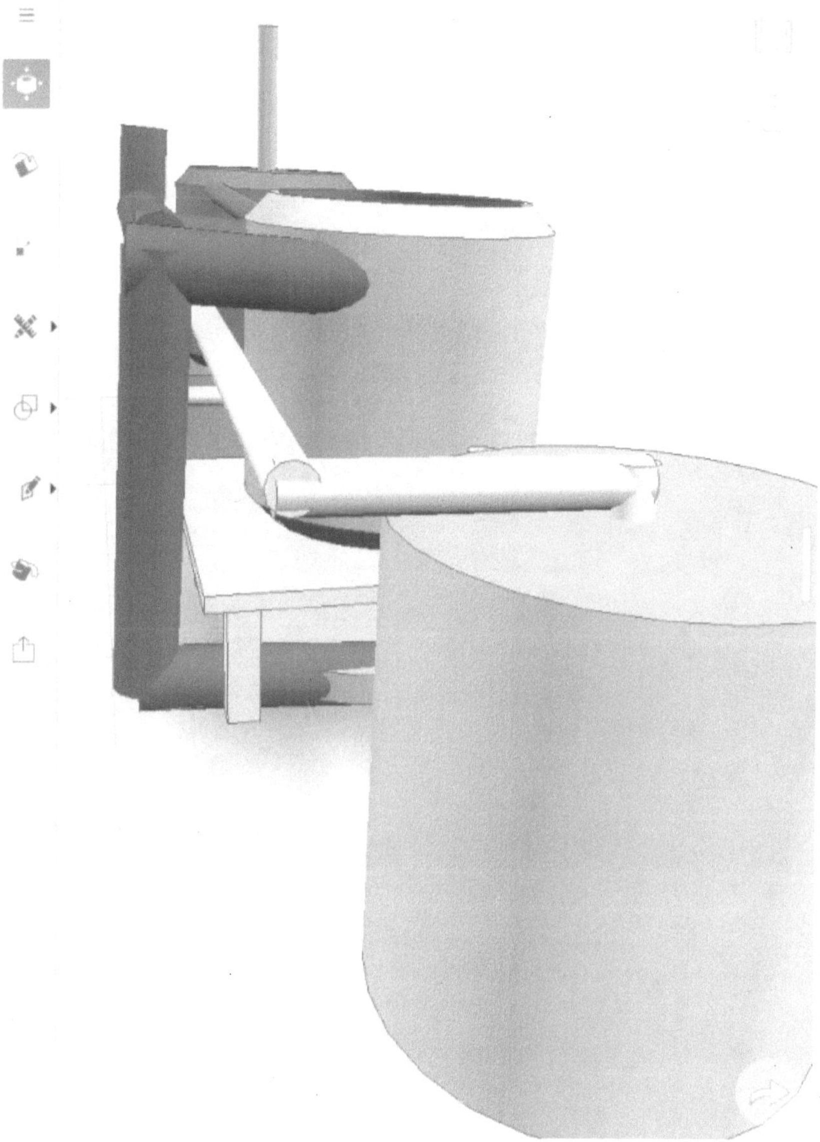

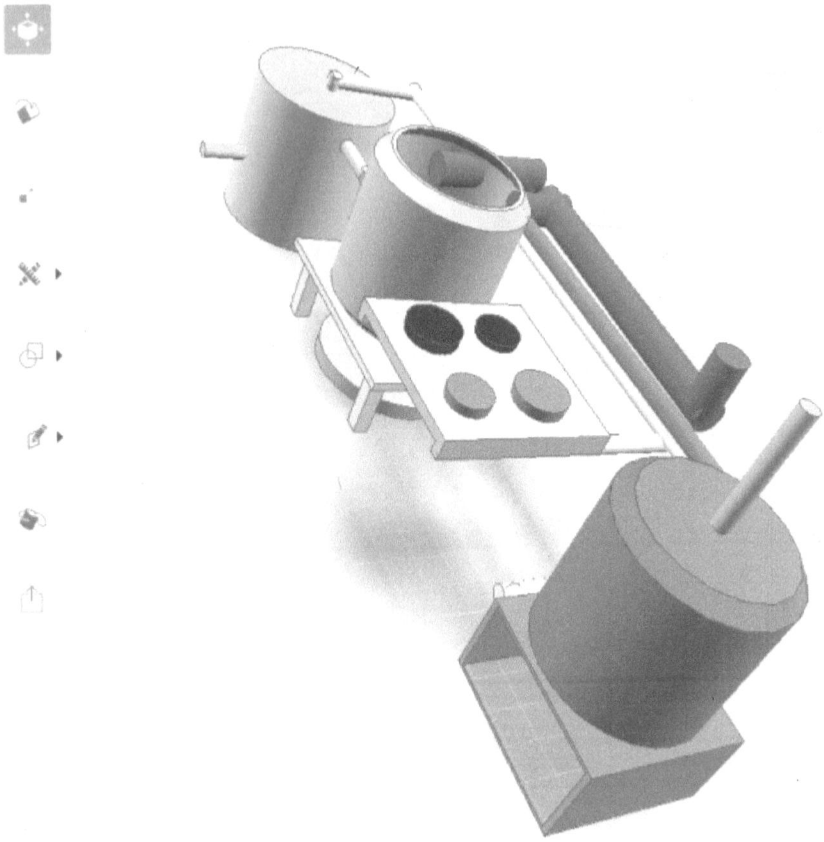

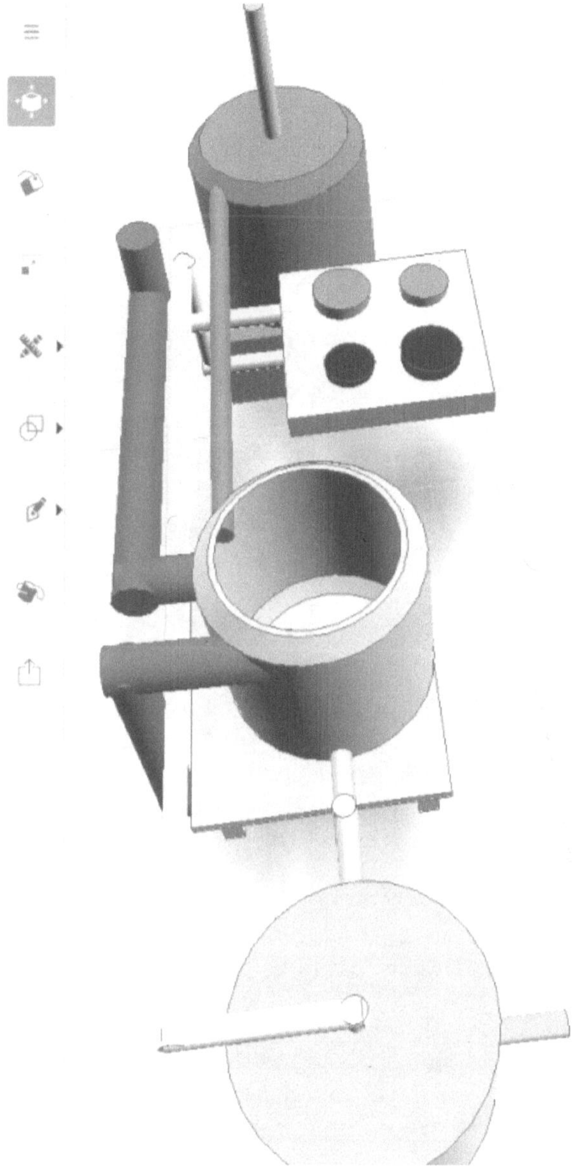

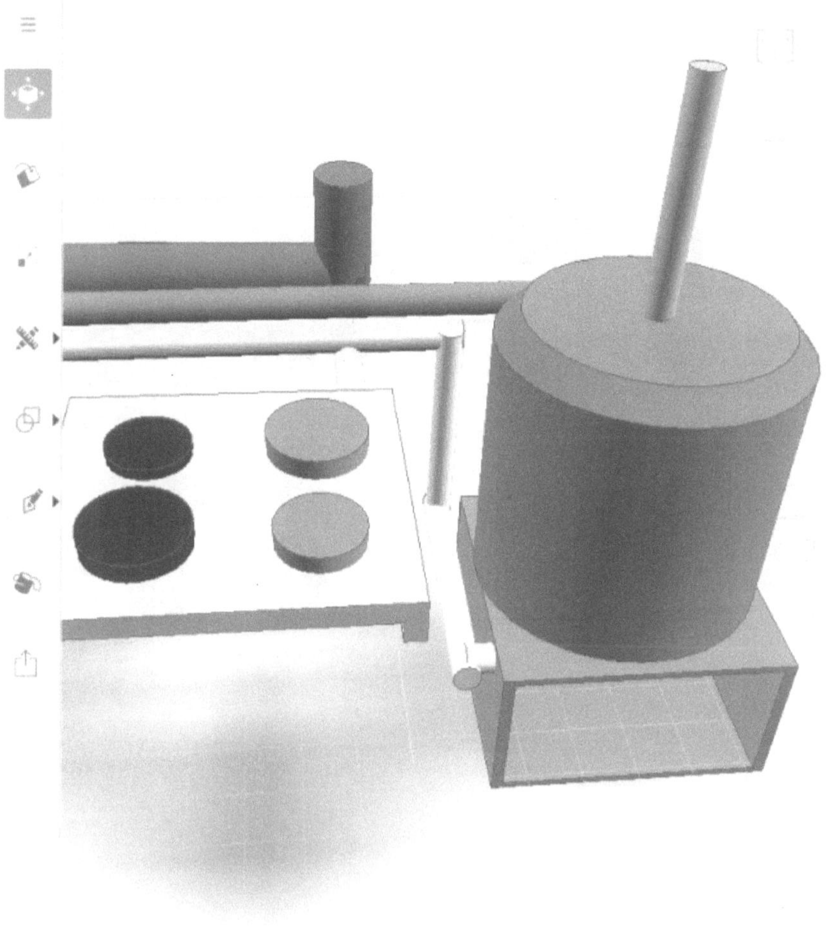

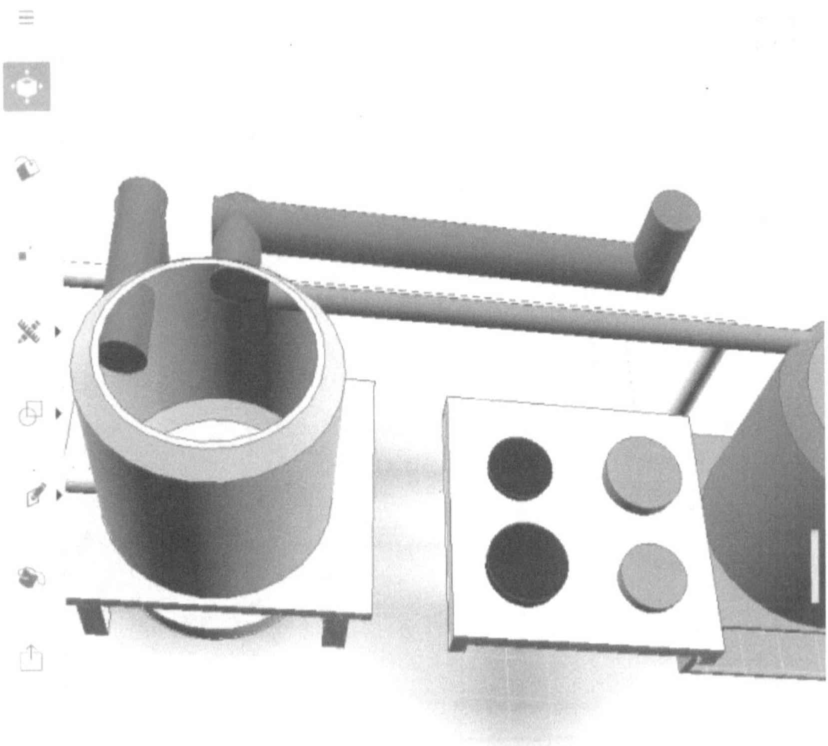

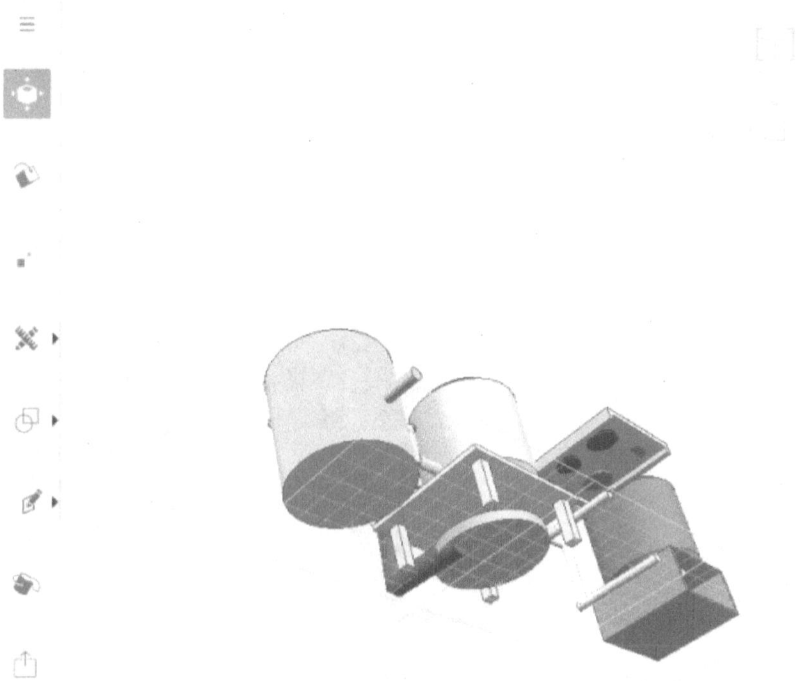

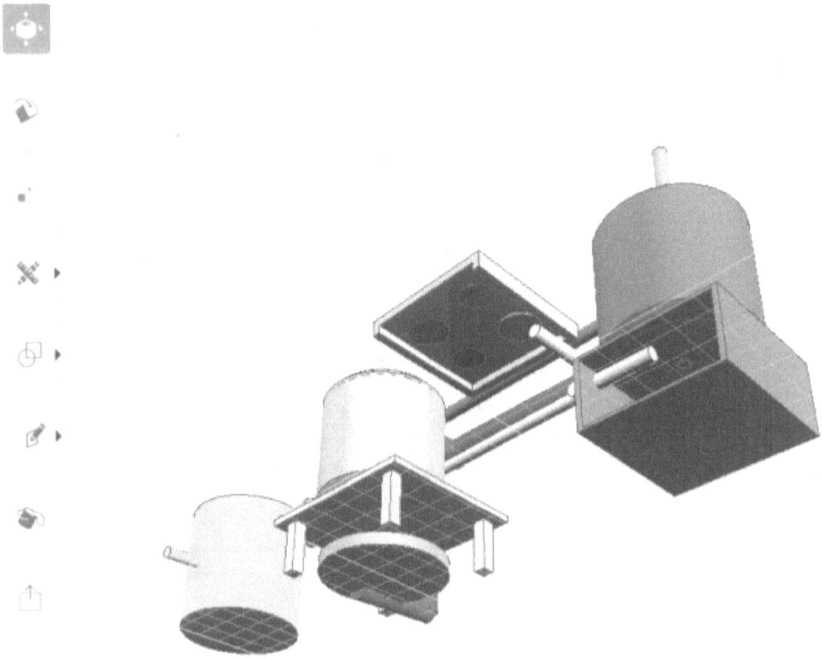

The washing machine (white square) and its tank (dark blue) should be installed 40 cm above the floor, in order to make it easier to dispose of the water with a faucet.

This sewage system (dark brown pipes) has a biodigester (green), sewage tank (light brown), where you decantate the solids and the liquids of your graywater, a ditch (orange), for the eventual excess of served water, boiling tank (purple), vapor (light blue pipe), a furnace (red square) and biogas (white pipes).

# Energy

Renewable energies include solar light, solar heat, wind, biomass, hydrogen and geothermal. Solar light, wind and geothermal may be too expensive, so we will focus on solar heat and biomass.

Parallel to the production of energy, it's necessary to use less energy to be sustainable. It means a house with good insulation, low-energy lamps, electronics and home appliances, and the minimum waste of energy (from fragile wiring system, heating water without need, etc.).

Any car can be adapted to use ethanol, but it's better to buy one with that technology already installed and remember that ethanol can be as expensive as gasoline, because it takes more volume to produce the same energy.

Electric cars are considered an appropriate choice (when they become cheaper), as well as bicycles (use of both of them is important). Using public transportation takes less resources from nature, but it can be more expensive, and it's not always available.

In conclusion, try to save (money AND energy) as much as possible when choosing your means of transportation.

Our ecohouse uses sun heat and biomass to produce energy. All domestic appliances use electric energy to work, but you can install a dual energy stove, that uses electricity in two burners and biogas in the other two. You can also use a firewood stove to cook, and in that case please remember to invite the author and his family for a meal.

Leftovers that cannot be sent to the biodigester shall be burned on the furnace or used directly on the soil, like paper, cardboard, feathers, coffee grounds, garden clipping, sawdust, plastics, chewing gum, nails, lint, hair, soil, sand or shells.

The vapor from the heating system is used to generate energy, moving an **2kW 2800 rpm 240 V monophasic electric motor with a turbine or similar**, installed on the third floor. In order to ensure energetic safety, it's important to have two generators, plus one for the pressure from vapor originated at the greywater and the furnace. The vapor will condensate and return to the water tank. Use a faucet to control the amount of vapor going to the turbine, and a manometer to measure the pressure. You can also use a controller, an inverter and some batteries, especially if you are also using photovoltaic panels.

冷水タンク
太陽熱コレクター
蒸気発生器
（ボイラー）
クラックトーム
KRAK-THOOM
ウンルコ
冷水タンク
蒸気圧タービン
発電機
蒸気発生器
（ボイラー）
冷水タンク

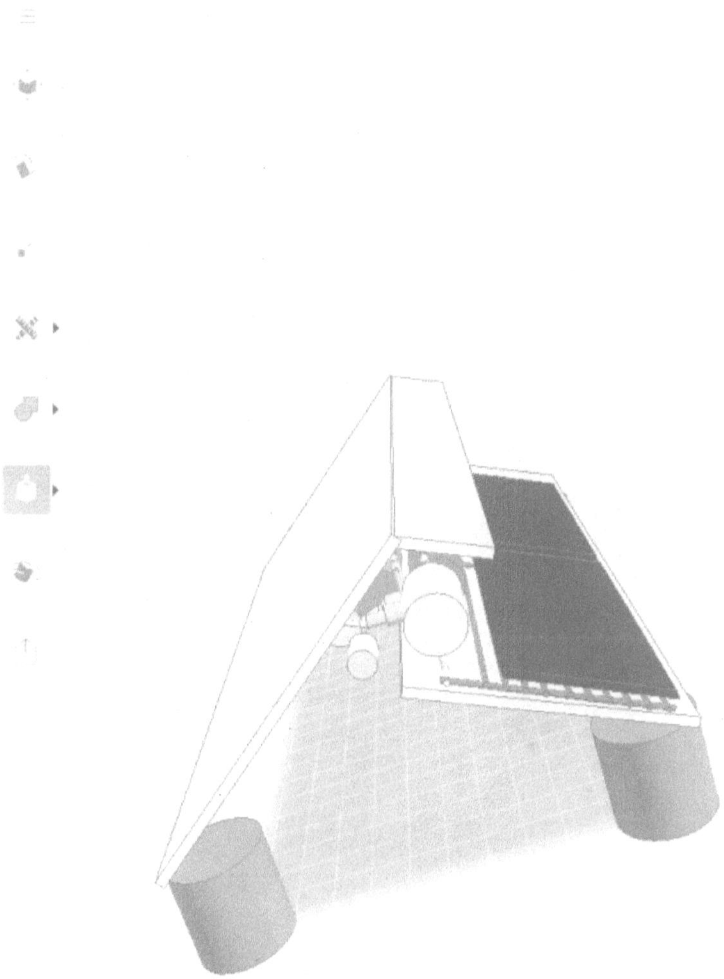

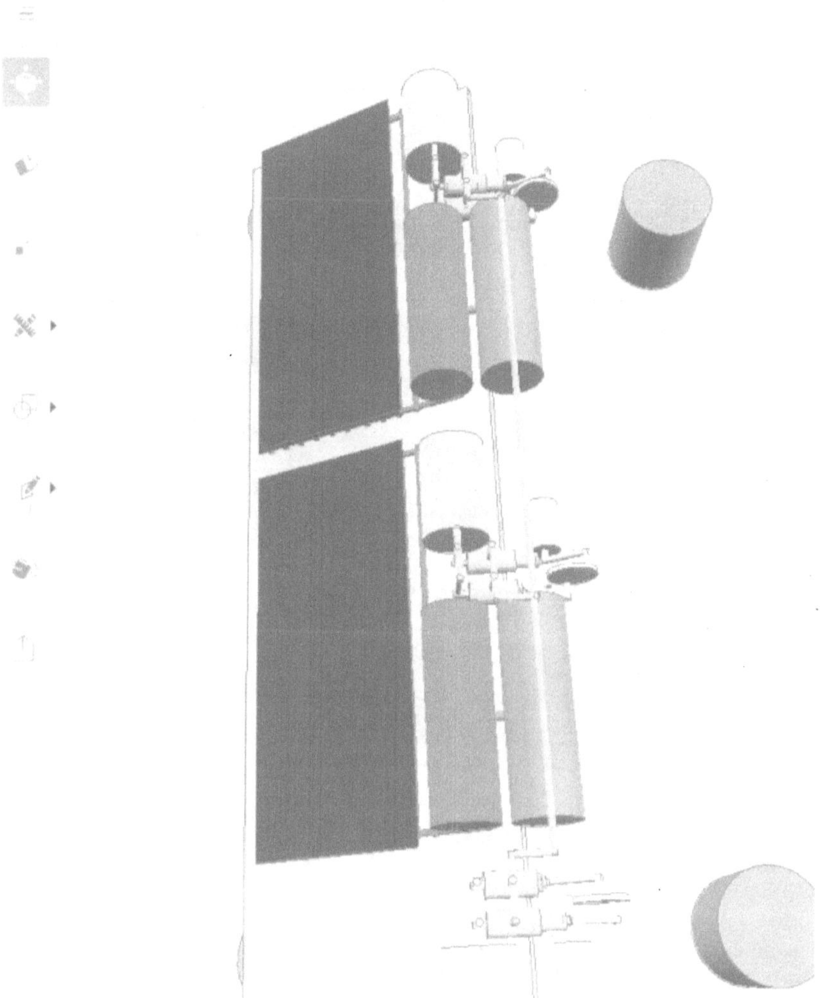

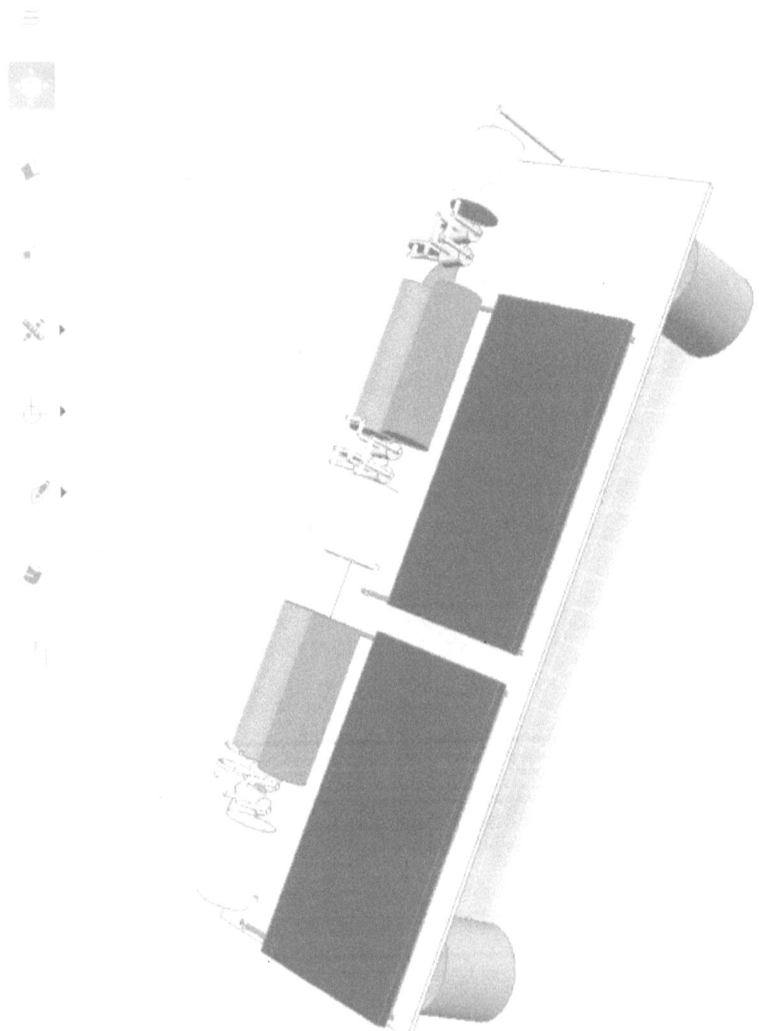

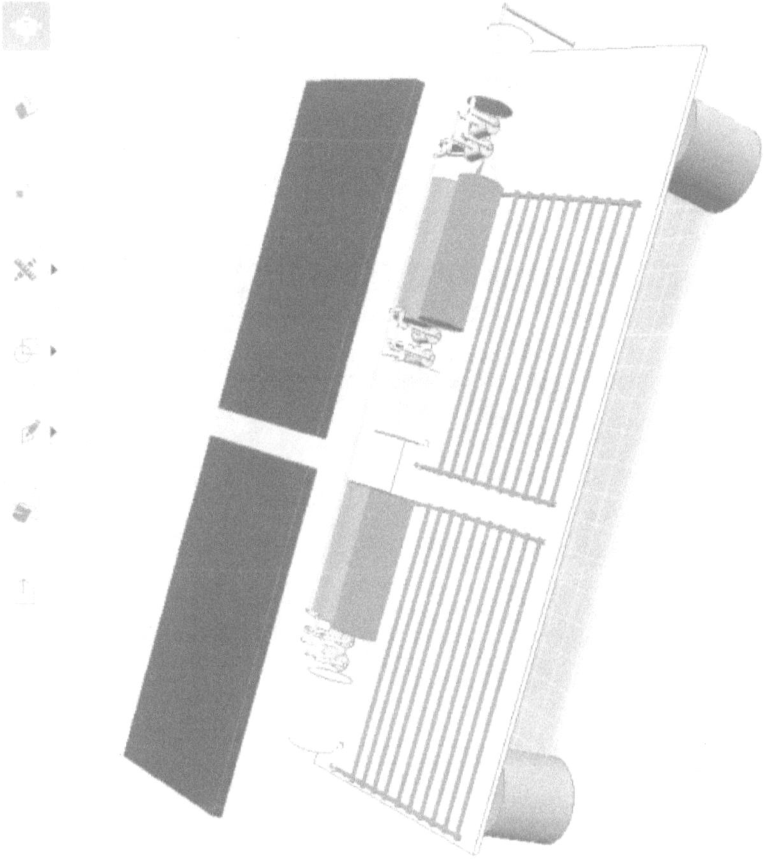

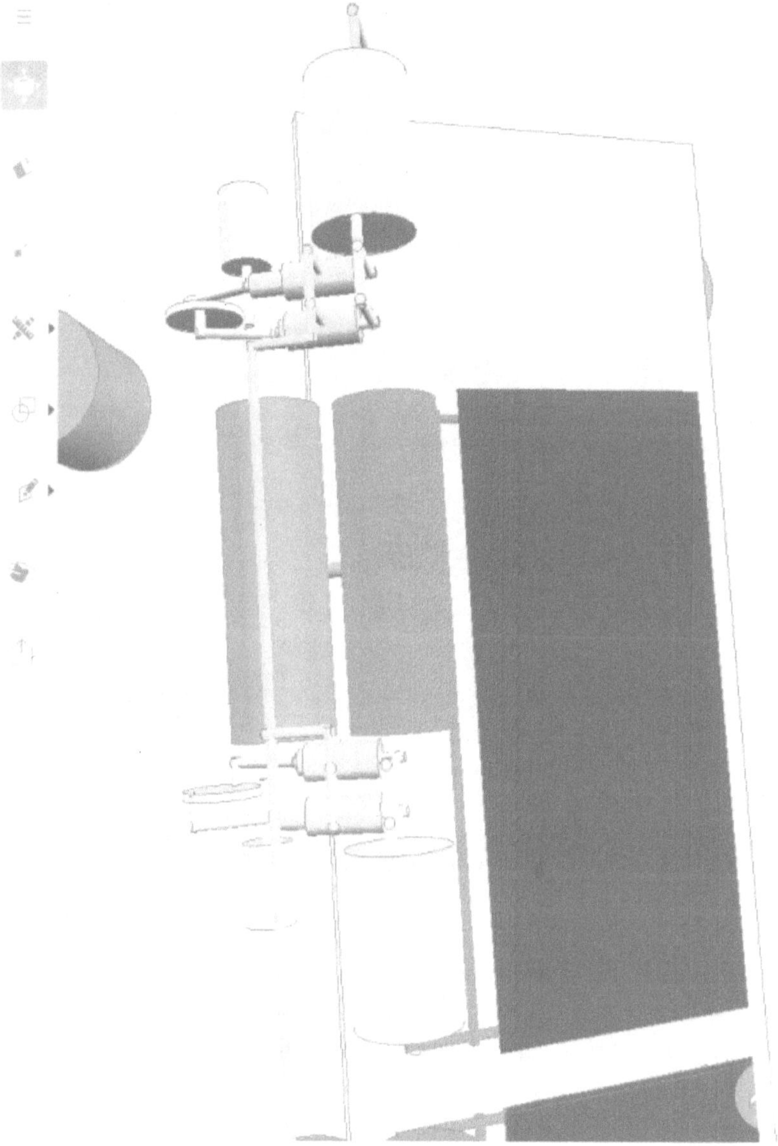

The system exposed above can heat water to temperatures above 100°, something common sun heaters cannot achieve. Over the metal pipes with water (in red) you need to install a black metal plate (black) and cover the empty spaces between pipes with polystyrene (Styrofoam).

The metal plate absorbs the heat, which is concentrated through the Styrofoam and makes the water in the pipes to boil. Use copper pipes or another metal that can withstand high temperature and pressure. The steam is concentrated in a proper tank (orange) and sent continuously to the steam engine (green). The steam moves the electric engine (gray) and is sent back to the water tank (blue). This is NOT a *moto perpetuo* machine, on cloudy and rainy days you need to use the furnace to generate electricity. During the winter the trees are dryer, and it's necessary to burn more firewood at the furnace in order to produce the same amount of energy.

Since you have enough yard for large trees, you have a good supply and natural storage of chemical energy, and you need to chop the branches and let them dry for optimal use in the furnace. The water boiled with the furnace (light blue) comes to a steam engine and after moving the electric generator it is sent to the water tanks.

You can install the solar panels on the upper roof, using a structure above it facing the sun (don't cover the sun heat panels and don't install it directly on the upper roof, it requires good ventilation and it is against the sun) at an angle equal to your latitude for optimal conditions. Remember it's necessary to use an inverter to transform it to 110 / 220 V alternate current, and diodes to prevent one from burning the others.

In a house energy is mainly spent on the following appliances: electric shower, air-conditioning, ironing, stove/oven (electric or gas), washing machine with dryer, and illumination. Avoid the first two (use some of the hot water to shower, and a good insulation system like the one shown before), and use low-energy appliances (LED lightbulbs, washing machine and dishwasher with no dryer, and so on).

# Food

Access to food depends mainly on financial resources, and to achieve that a person needs to work, and to spend less money on other things.

The ecohouse has about 50 sq. m. on the roof to be used for planting edible plants, and an infinite amount of fertilizer from the water-recycling system (biodigestate). Use the first floor to plant fruit trees, and the third floor to plant other edible plants (tomatoes, watermelons, carrots, potatoes, cabbage and arugula are plenty of vitamins and minerals, and easy to grow). Breeding animals like cows, chickens and others should be done away from the house, in order to prevent sanitary issues.

Much money is saved using sustainable energy and water sources. Work with making bamboo furniture and domestic utensils is also a non-negligeable source of revenue. Also, working on building ecohouses, adapting and making improvements on it, can also be profitable, and can be shared for common use of mankind.

If you live in a rural area, plan very well where the plants will be placed, those that require daily care should be close to the house, while crops and large trees can be located in more distant areas. In this case, build your home in the center of the property, to avoid leaving any area too far away and being forgotten over time. More information on the subject in the book *The Self-Sufficient Life*, by John Seymour.

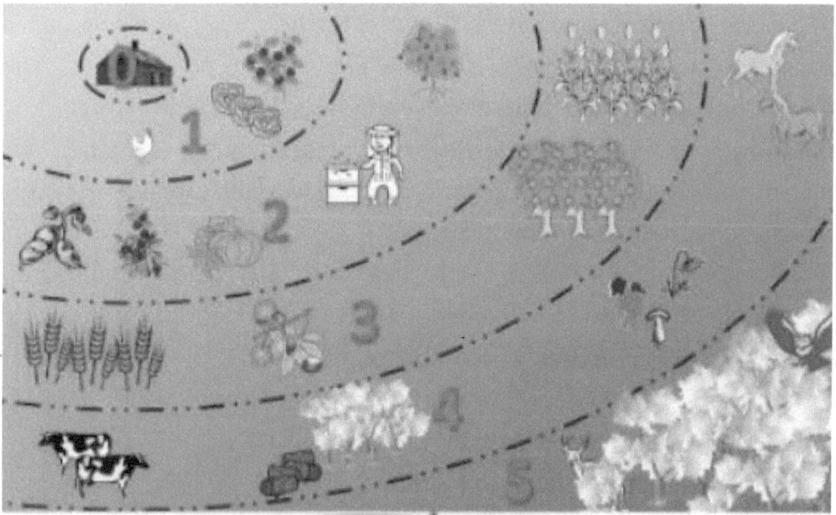

Zone 0 (The House): The central hub where people live, cook, and plan the system.

Zone 1 (Daily Use): Immediately adjacent to the house. Contains herb gardens, sprouts, seedlings, medicinal plants, rainwater harvesting systems, and household composting. Everything you harvest or tend to every single day.

Zone 2 (Semi-Intensive): Slightly further away. Includes small orchards, perennial plants, chicken coops (for pest control and fertilizing), small ponds, and main vegetable beds.

Zone 3 (Large-Scale Crops): Pasture areas for larger animals (cows, sheep) and large field crops that do not require daily care, such as corn, wheat, root crops, or commercial crops.

Zone 4 (Semi-Wild): An area for silviculture and forest management. Used for harvesting firewood, stakes, wild fruits, and native plants. Requires very little intervention.

Zone 5 (Wild Nature): An untouched ecosystem where humans enter only to observe, meditate, and learn from natural intelligence. There is no design or cultivation here.

An intelligent way to take advantage of nature in your favor is to place plants that fix nitrogen in the soil, such as beans and soybeans, alongside other cereals such as corn and oats, avoiding the wear and tear and consumption of fertilizers, in addition to the consecrated crop rotation. Remember that the animals you raise also need food, so you should plant what they will eat, something more sustainable than buying their food.

Animals on the property can be a blessing or a nuisance in the resident's life. Insects such as ants, termites, caterpillars and aphids need to be fought by putting ashes on the feet of fruit trees, planting tobacco in the vegetable gardens (the third floor of the ecohouse is well protected from some pests, easing the production) or water from the washing machine.

Pigs, goats and chickens can be raised on plots together with crops (but on separate plots, of course). These animals clean the soil (chickens enter last, to remove weeds) and fertilize the land in the plots where they are confined, doing the work for the planter and placing their waste where they are needed. In addition, you won't need to harvest what you've planted and leave it stocked to give to the animals, they do this service themselves when changing the plots, saving time, effort and resources.

No doubt the manual work of planting, watering, taking away the weeds, fertilize and harvest will require time and effort of the resident, but all this work shall be shared among the house dwellers, especially the younger ones. In order to learn how to valorize and respect the environment it is essential to have direct contact with it and reap its fruits.

Lastly, we emphasize that we don't want any profits from these ideas. Our only goal is to improve the well-being of people in need.

*Thank you for your effort in making Earth a better place to live.*

*God bless you.*

###

This book is the opinion of the author, and nothing else; it does not represent the opinion of any government, organization or any third parties.

Thank you for your interest in reading this book. I really appreciated that.

Certainly, many people will not agree with it, as it's common in any scientific discussion... So, I would like to know your point of view.

Feel free to send suggestions, comments and opinions to rogeriocietto@gmail.com, subject Ecohouse. Your email is very welcome.

I regret to inform that you will not find me on Facebook, Twitter, Orkut or any related media.

www.ingramcontent.com/pod-product-compliance
Lightning Source LLC
LaVergne TN
LVHW090430210926
848480LV00002B/762
* 9 7 9 8 2 3 3 0 1 8 8 3 1 *